THE U.S. PATENT OFFICE IS
MORE THAN 200,000
APPLICATIONS A YEAR.
SOME OF THEM ARE WORKS OF PURE GENIUS.
OR PURE INSANITY.
DON'T MISS . . .

- The Breath Detector—Never be embarrassed again by bad breath. This strange contraption lets you smell your own breath before a romantic encounter.

- Eye Protector for Chickens—Don't expose your favorite hen to needless pecking dangers. These tiny goggles are more than a fashion statement!

- Apparatus for Exercising the Hearing Organs—When you want a total body workout, don't forget ear calisthenics. Keep your eardrums toned and flab-free!

- The Kane Golf Ball—Has losing golf balls become a bit expensive? This innovative design for a smoking ball marks your lie . . . or sets the greens on fire.

- The Mey Velocipede—You'll never worry about gas prices going up when you use dog power to run your car! And you'll never get another speeding ticket either.

- Weapon—It only looks like a hat. This devious device turns your head into the mount for a mini-machine gun. But the wearer might be in for a hell of a migraine.

MR. McMURTRY'S

BUBBLE HAT

. . . AND OTHER GREAT MOMENTS IN AMERICAN INGENUITY

Mike Miller

A Dell Trade Paperback

A DELL TRADE PAPERBACK

Published by
Dell Publishing
a division of
Bantam Doubleday Dell Publishing Group, Inc.
1540 Broadway
New York, New York 10036

Library of Congress Cataloging in Publication Data

Miller, Mike (Michael W.), 1962–
Mr. McMurtry's bubble hat—and other great moments in American ingenuity / Mike Miller.
p.   cm.
ISBN 0-440-50657-3
1. Inventions—United States—History.   2. Patents—United States—History.   I. Title.
T21.M57   1996
609.73—dc20   95-54179
                                                                  CIP

Printed in the United States of America

Published simultaneously in Canada

September 1996

10   9   8   7   6   5   4   3   2   1

FFG

INTRODUCTION

Not long ago a man named Rick Carballo traveled to an ultramodern suburb in Virginia just across the highway from Washington's National Airport. He made his way into a sprawling office park crammed with forbidding towers. His ultimate destination was a glass and steel honeycomb called Crystal Plaza Three. This, he knew, was where he would find all there is to know about condom-pocket underwear.

He made his way into a cavern of metal shelves piled high with neat stacks of loose paper. They were organized in an elaborate taxonomy that seemed to encompass every nonliving object on earth, as well as some living ones.

One corner had the section Mr. Carballo was looking for: "Class 2: APPAREL." He zeroed in on subclass number 2, "Guards and Protectors." It had dozens of subdivisions of its own: Fireman's Helmets, Armpit Shields, Try-On Hat Linings, Burial Garments, Necktie-Engaging Devices, Head Coverings (including subsubdivision number 04, "Tam o' Shanter Type"), Adornment Handkerchiefs, and many more.

Finally he found subdivision 400: Underwear.

Like many before him, Rick Carballo had come here hoping to find only a meager selection of drab designs, leaving plenty of room for improvement. But he was disappointed. Before him was a staggering feast of underwear ingenuity: underwear with pouches for ice, underwear with removable moistened towelettes, underwear with slideable crotch panels—and, number 5,172,430, condom-pocket underwear, issued on December 22, 1992, to Natalie A. Lerma-Solis of Stockton, California.

Thomas Edison, who said genius is 1 percent inspiration and 99 percent perspiration, left out the most important ingredient of all: innovation. The wizard who shouts "Eureka!" and bursts from his lab with flawless drawings for a lightbulb that looks just like Tom's has nothing. His idiot cousin who belches and falls off his barstool with a doodle on a napkin for condom-pocket underwear has a patent.

That's because, deluged with more than 200,000 applications a year, the U.S. government generally grants patents only to inventions it finds to be truly novel. On the question of whether they are truly useful, the government displays magnificent latitude, as you will discover in this book.

Thus the most pressing question on the minds of hopeful inventors is: Did I just spend a million hours of all-nighters in my garage unwittingly retracing some other wizard's (or idiot's) steps?

So from across the country they flock to Crystal Plaza. There they can scour the U.S. Patent Office's Public Search Room, official repository for all 5.5 million patents ever granted, beginning with the first, issued on July 31, 1790, to Samuel Hopkins of Pittsford, Vermont, for a recipe for potash and pearl ash (crucial eighteenth-century ingredients for soap).

On their way into the Search Room, the hopeful inventors all pass a loving shrine to great moments in patent issuing. There, proudly on display behind glass, are number 174,465 ("Telephone"), number 1,647 ("Telegraph"), number 223,898 ("An Incandescent Lamp"), and a cavalcade of others.

This informal Inventor's Hall of Fame offers a vivid reminder that every so often a gadget gets a patent and changes the world. (Or the language, as in patent number 608,845, "Internal Combustion Engine," issued to Rudolph Diesel.)

Some of the inventions here fill such simple needs that it's easy to imagine somebody daydreaming about them. Take patent number 821,393: "A Flying Machine."

Others sound far too bizarre to interest anyone, much less wreak a revolution. Patent number 2,543,181 went to "Photographic Product Comprising a Rupturable Container Carrying a Photographic Processing Liquid." Its inventor, Edwin Herbert Land, later thought of a catchier name: Polaroid.

But this book isn't about the hall of famers. It's a reminder that in the history of American genius, for every Thomas Alva Edison, there have been a million Alden McMurtrys.

Mr. McMurtry was the Connecticut tinkerer who in 1911 rushed to the U.S. Patent Office with his immortal design for the bubble hat. It used a hidden gas canister to send soap bubbles out of a hat—perfect, Mr. McMurtry thought, for showstopping chorus numbers. It looked reasonable to the official government examiners, who handed Mr. McMurtry patent number 1,021,323.

On the shelves of Crystal Plaza Three, there are far more bubble hats than incandescent lamps. There are housemaids' shoes designed especially for pumping bellows, a rocking chair that churns butter, a menacing hat that fires bullets when the wearer blows into a tube, and a handy pair of eye protectors for chickens. All these, and a myriad of other awesomely impractical concoctions, are in the pages that follow: a gallery of great McMurtryesque moments in American invention.

Today the McMurtrys of the world are sober persons of commerce. They have lawyers and professional patent searchers, whom they hire to comb through the Crystal Plaza stacks—sleuths like Rick Carballo.

"We look for everything and anything," said Mr. Carballo, a small mustached man in a V-neck sweater who recently paused in mid-hunt to explain his unusual trade.

"It's like playing chess in a lot of ways. You find one patent, and if it resembles what you're looking for, you use its references to find something closer. So you keep moving from one patent to another until you find the duplicate of your invention."

Finding a duplicate is what dashed the hopes of his client with a vision for underwear with a little pocket for a condom. (The discreet searcher wouldn't identify the thwarted inventor.) "They call us dreambusters, because that's what we do," Mr. Carballo said.

Another searcher sauntered by. "What are you hunting on?" he called out.

"Covers for wheels while you paint the car," Mr. Carballo replied.

The other searcher, a Crystal Plaza old-timer, nodded knowingly. "I'd have to say there's a spot for that in one fifty," he said. " 'Protective Covers.' "

Still, in the Search Room crowd, there are a few genuine McMurtrys foraging about the shelves.

"Excuse me, where would I find golf clubs?" A dapper elderly gentleman in a tweed jacket and flannel slacks nodded politely as he received directions to the Amusement Devices section. He was Guy Parr, a retired Maryland laboratory administrator who recently had a thunderbolt of insight about putters.

Suppose, Mr. Parr found himself musing, the handle of a golf putter joined the head at its center instead of at one end. It could swivel back and forth from one end to another. Thus right-handers and left-handers could use the same putter!

Mr. Parr quickly realized his putter also had exciting implications for chippers. Because their heads jut out at an angle, chippers also come in separate right-handed and left-handed models. But what if the Parr putter used its back side as a chipper? Then you'd have four clubs for the price of one (righty putter, lefty putter, righty chipper, lefty chipper, depending on how you held the club and where you moved the swivel). Before long he was racing off to Crystal Plaza Three.

It wasn't his first trip. Three years earlier the U.S. Patent Office recognized Mr. Parr and a partner as the inventors of the microwave food cover. "It's just a wax paper cone," he said. "You say to yourself, 'Is someone going to reinvent the wheel and patent a cone?' But sure enough, we got the patent on the cone."

Mr. Parr took his cone around to a few big players in the foil and wrap industry. But nobody beat a path to his door. He has higher hopes for the ambidextrous swivel putter-chipper.

"It's like the lottery. You spend your money, and if you win, you might win big," he said. "Some ideas do it—like the Abplanalp valve for the spray can. He was the man that Nixon used to stay with. That valve made him fabulously rich. But his idea was deadly simple. You just think of something that simple—and you have the great American dream."

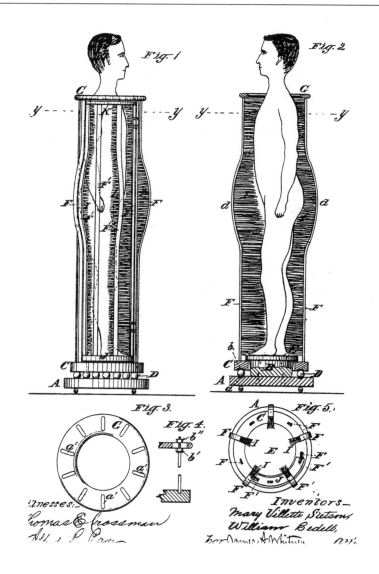

Fig. 1

Fig. 2

Fig. 3.

Fig. 4.

Fig. 5.

Witnesses:
Thomas E. Grossman

Inventors—
Mary Villette Sutton
William Bedell,

FLESH BRUSHING APPARATUS
Patented June 20, 1882
by MARY V. STETSON AND WILLIAM BEDELL
New York, New York

Flesh brushing was never quicker and easier than it was with the Stetson/Bedell apparatus. Instead of laboriously brushing all your flesh by hand (and missing lots of hard-to-reach places), you simply climbed into a big tube of "sea-root" bristles. Then you grabbed a pole and rotated the bristles around your body, "thereby effecting a great saving of time and of exertion in this operation of the toilet."

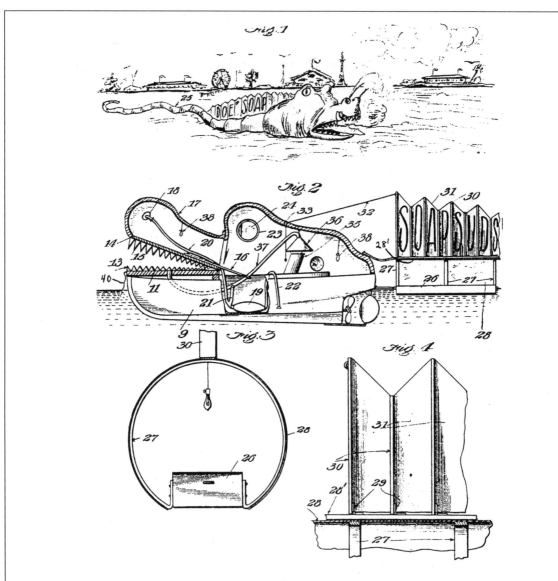

ADVERTISING DEVICE
Patented October 31, 1905
by MAXWELL S. ALEXANDER
Los Angeles, California

Mr. Alexander earned his footnote in marketing history by dreaming up an extremely specific new location for an ad: an artificial sea serpent fin.

A pilot would sit inside the serpent's head, flashing lights through the eyes, shooting water through the nostrils, and wiggling the ad "as the serpent is supposed to raise or lower its fin." The effect, Mr. Alexander confidently predicted, would "call the attention of all beholders to the float and the advertising matter thereon."

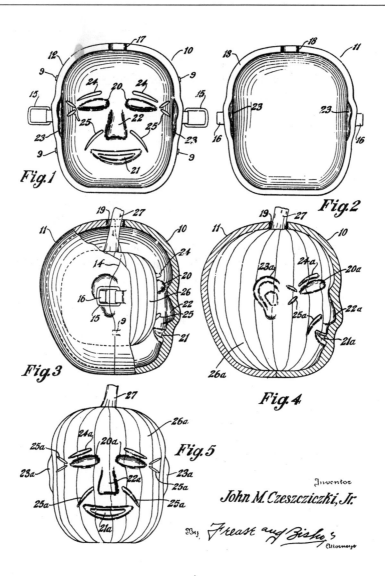

Fig.1

Fig.2

Fig.3

Fig.4

Fig.5

Inventor

John M. Czeszcziczki, Jr.

By Frease and Bishop
Attorneys

• 4 •

FORMING CONFIGURATIONS ON NATURAL GROWTHS
Patented October 19, 1937
by JOHN M. CZESZCZICZKI, JR.
Madison, Ohio

Mr. Czeszcziczki clearly never enjoyed happy boyhood memories of carving pumpkins at Halloween. He proposed a technique for manufacturing precarved jack-o'-lanterns right on the vine, by popping immature pumpkins into a face-shaped mold.

The unsentimental inventor noted that his jack-o'-lanterns would require no messy cutting and no "marring the shell." It would also be more reliable than the old Halloween tradition, in which, he grumbled, "considerable skill is required to produce a good likeness."

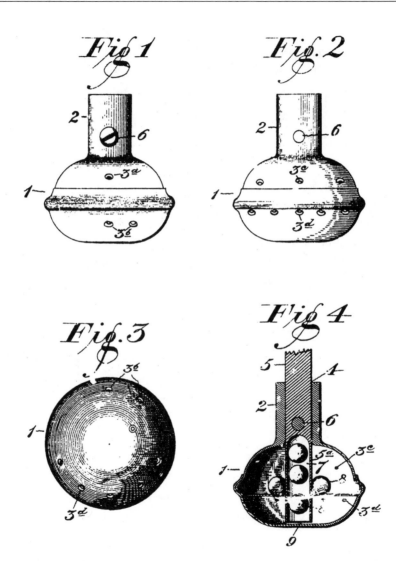

Fig.1 Fig.2 Fig.3 Fig.4

SANITARY DOORKNOB
Patented November 25, 1930
by BASILIO POLCARI
Somerville, Massachusetts

For Mr. Polcari, the humble doorknob was a pestilent hot zone of contagious disease. In a terrifying preamble to his patent, he explained: "It is easy for the hands to become infected with poison oak, erysipelas, itch, leprosy, etc. When persons having contagious diseases in their hands take hold of door knobs, the disease germs are left there in position to be taken up by the hand of the next person who comes and takes hold of the knob. And thus disease is spread."

Mr. Polcari set out to stem this epidemic with a hollow knob filled with little balls of disinfectant. As the knob turned, the cleanser would seep out through tiny holes onto the thumb and fingers.

He also cunningly enlisted an unwitting ally in his war against disease: "When the knob is brightened up by the housekeeper, the disinfectant coming out through the perforations will be caught and applied all over the outer surface."

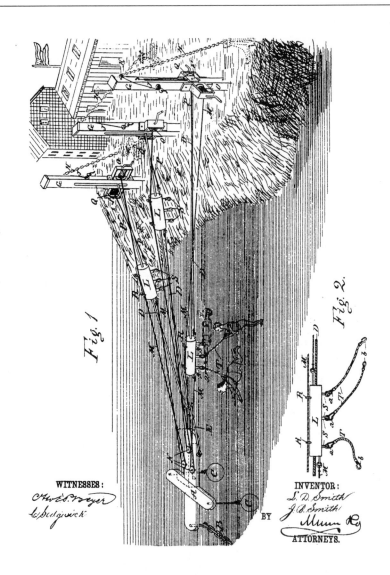

Fig. 1

Fig. 2.

INVENTOR:

L. D. Smith

J. D. Smith

BY *Munn & Co*

ATTORNEYS.

SAFETY APPARATUS FOR SEA BATHERS
Patented August 15, 1882
by LORENZO D. SMITH AND JOHN B. SMITH
Patchogue, New York

With the Smith safety device, swimmers could take a carefree ocean dip even in the choppiest waters.

They simply strapped on a harness and hooked themselves up like cable cars on a long rope. On a busy day the beach would become a whirring little assembly line of swimmers gliding back and forth on their cables, with lifeguards reeling them in whenever the undertow became too strong.

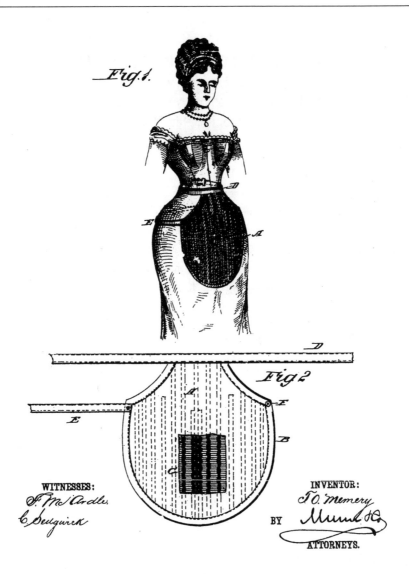

Fig. 1.

Fig. 2.

WITNESSES:

F. McArdle.
C. Sedgwick.

INVENTOR:

T. O. Memery

BY

Munn & Co

ATTORNEYS.

DRESS GUARD
Patented July 13, 1880
by Tom O. Memery
Key West, Florida

The discreet Mr. Memery sought to design a ladies' accessory to "prevent the dress from clinging so very close to the body, as it does in stormy and windy weather, and thus prevent the form from being exposed."

His handy solution: an apron "stiffened by a series of strips" of steel or whalebone. He also delicately proposed an extradense patch of strips in the middle of the apron, "so as to give greater stiffness to that part."

FIRE-ESCAPE
Patented November 18, 1879
by Benjamin B. Oppenheimer
Trenton, Tennessee

Mr. Oppenheimer boldly promised a safe, easy escape from "a burning building from any height."

As the flames came rushing toward you, you'd simply throw on a pair of enormous shoes, strap an awning ("stiffened by a suitable frame") onto your head, and leap out the window. Thick elastic pads on the soles would "take up the concussion with the ground," and you'd land "without injury and without the slightest damage."

Fig. 1.

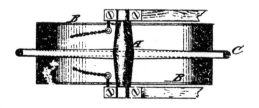

Fig. 2.

VELOCIPEDE
Patented November 29, 1870
by F. H. C. MEY
Buffalo, New York

Why pay for fuel if you have a frisky dog just bursting with unharnessed energy?

That was the revelatory insight behind the Mey Velocipede, or "Dog Power Vehicle." Your dog climbs inside a "treading-rim," like a gerbil in an exercise wheel, and away you go!

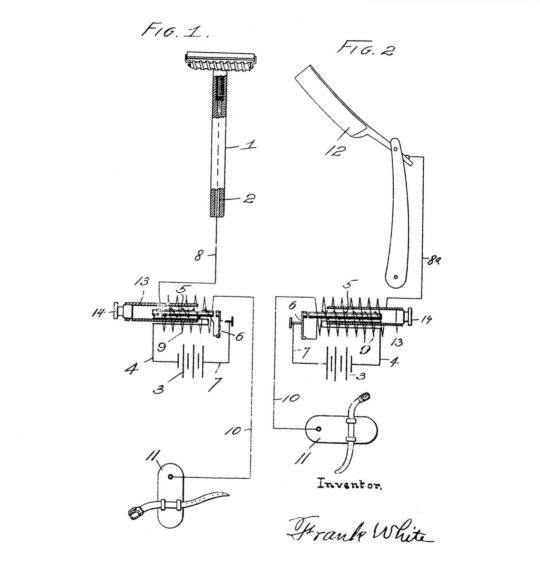

FIG. 1.

FIG. 2

Inventor.

Frank White

RAZOR
Patented January 14, 1919
by FRANK WHITE
St. Louis, Missouri

Mr. White had a good idea—an electric razor—but he didn't quite get the concept right.

His somewhat unsettling idea was to hook an ordinary hand razor up to a battery and send a "mild" electric current to the shaver's face. The nifty result: a good zap in the cheek, producing "the effect of raising the ends of the hairs so that they may be readily engaged by the edge of the blade."

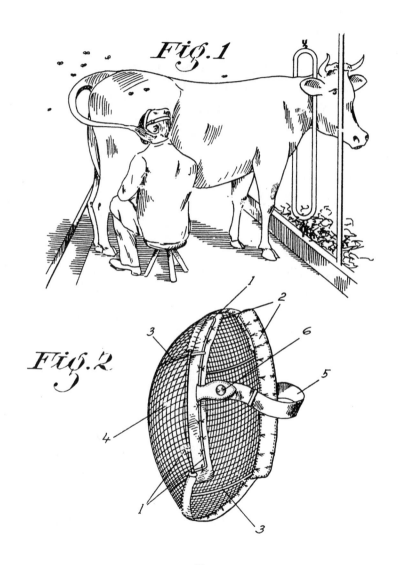

Fig.1

Fig.2

MILKER'S MASK
Patented January 7, 1919
by FREDERIC W. ELLEBY
Modesto, California

"he lashing of the face and eyes by a cow's tail, especially during fly season, is most obnoxious," Mr. Elleby observed. "Many different tricks are resorted to by a milker to avoid the same. None of the present day methods are satisfactory."

He pointed out that a baseball catcher's mask wouldn't do the trick because the bars are too far apart. But his Milker's Mask would be just perfect: openings large enough "to permit of easy vision therethrough" but small enough "to prevent any of the hairs of the animal's tail from working therethrough."

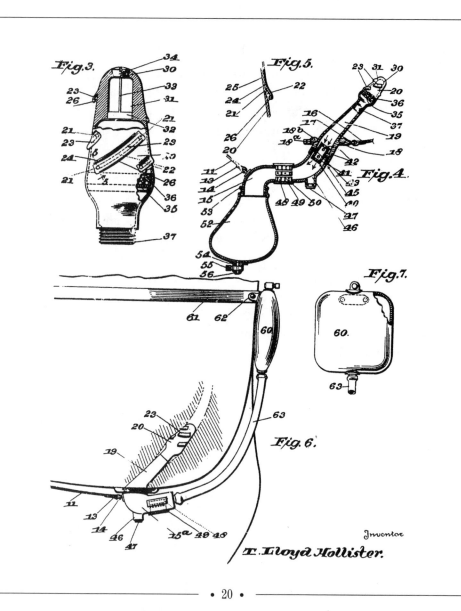

Fig.3.

Fig.5.

Fig.4.

Fig.7.

Fig.6.

Inventor

T. Lloyd Hollister.

GAS RECEPTOR
Patented February 28, 1939
by THOMAS LLOYD HOLLISTER
Miami, Florida

Exactly what gas did Mr. Hollister have in mind? "Gas formed in the alimentary tract of the body."

His unusual device featured a long "nipple" attached to an elaborate system of tubes, valves, and flanges, ending in a small bag. The nipple, he daintily explained, was "adapted to be received at the lower end of the alimentary canal." The fearless user would strap the gadget on with a belt around the waist and presumably feel much better when it was all over.

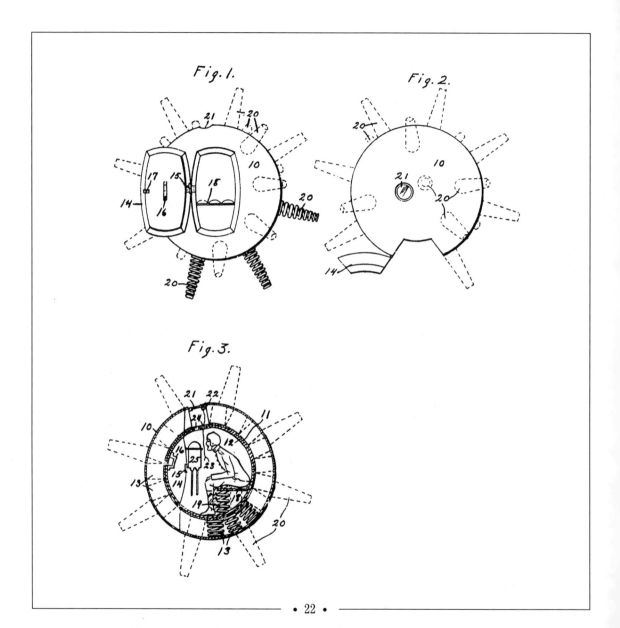

Fig. 1.

Fig. 2.

Fig. 3.

AIRCRAFT LIFE BOAT
Patented October 15, 1918
by TONY SALARI
Bisbee, Arizona

A ir travel today would be rather different if the Salari life boat had caught on. Each passenger would take his seat in a comfortable private sphere and seal himself shut with a refrigerator-style door. The sphere would be encased in springs, an outer shell, and more springs radiating out like sea urchin spines.

In the unlikely event of an emergency landing, all those springs would "resist a very great shock and provide against injury of a considerable fall." An extra bonus: a little window to watch the view on the way down.

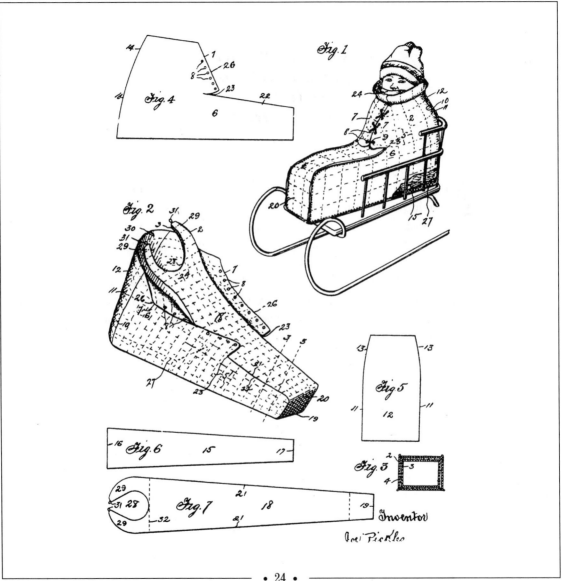

Fig. 1

Fig. 4

Fig. 2

Fig 5

Fig. 6

Fig. 3

Fig. 7

Inventor

Joei Pickles

• 24 •

PROTECTOR FOR CHILDREN
Patented May 7, 1918
by JOE PICKLES
Winnipeg, Canada

※

From the frozen north came this formidable defense against "the inclemency of the weather."

Mr. Pickles accurately observed that his invention "as a whole resembles a moccasin of very large size." A solicitous parent shoved the child inside and laced him up tightly, "thus inclosing the child up to the neck and preventing him from getting his arms out into the cold."

The Pickles protector protected against more than cold weather. It was lined with "a removable absorbent pad having a water proof bottom . . . the purpose of which is self evident."

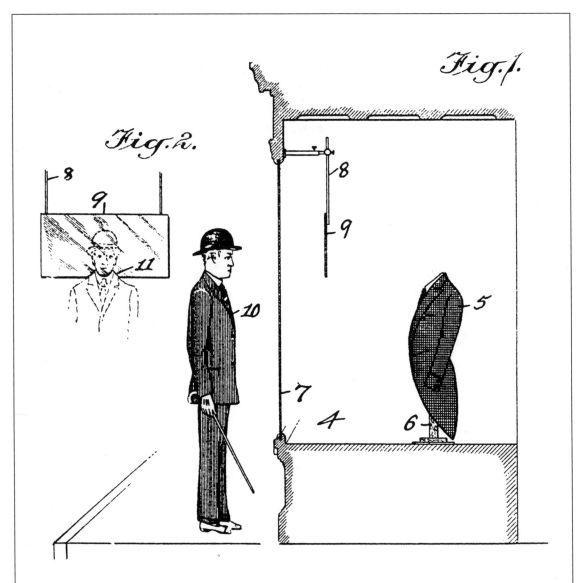

Fig. 1.

DISPLAY DEVICE
Patented August 7, 1917
by CHARLES MACQUESTEN
Bloomfield, New Jersey

For the ultimate window-shopping experience, Mr. MacQuesten designed a little head-sized storefront mirror with a chin-shaped curve cut out of the bottom. Standing in the proper position, a shopper "may view himself as in the act of wearing the article or articles on display."

"My device may be modified to meet many different conditions," Mr. MacQuesten noted. A few he didn't mention explicitly: cross-dressers, crash dieters, or anyone else fantasizing about a brand-new body.

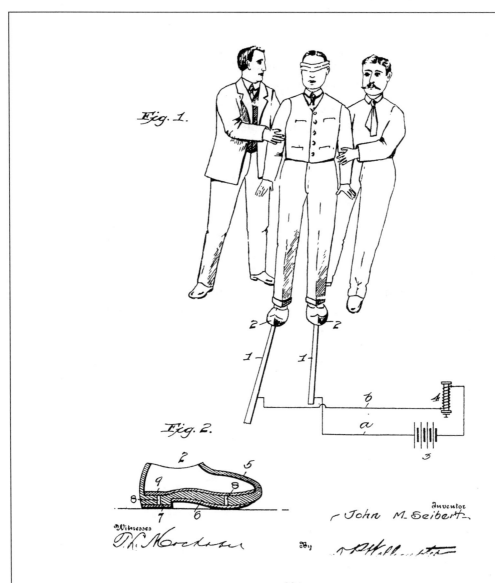

Fig. 1.

Fig. 2.

Inventor

John M. Seibert

Witnesses

T. L. Mockers

By

R. B. Hall...

INITIATION APPARATUS
Patented May 8, 1906
by JOHN M. SEIBERT
Pekin, Illinois

Mr. Seibert must have raised an eyebrow or two at the Patent Office when he insisted that his apparatus was "entirely harmless in its action and results, while at the same time producing an amusing and entertaining effect."

His invention was "for use in lodges and secret societies for initiating a candidate into the secrets and forms of the order." The eager candidate slipped on a pair of shoes with electrodes on the heels and soles. Then his lodgemates blindfolded him and led him along a pair of metal rails wired to an electrical generator.

The hilarious result: "The act of walking will of necessity open and close the electrical circuit, thereby causing the shocking of the subject."

FIG.1.

FIG.4.

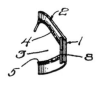

FIG.2.

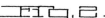

FIG.5.

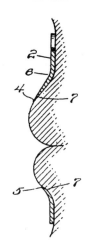

FIG.3.

Inventor
Roy E. Kirk

LIP MARKING GUIDE
Patented March 5, 1929
by ROY E. KIRK
Columbus, Ohio

With Mr. Kirk's thin sheet-metal stencil shoved against your mouth, applying lipstick was a breeze, "forming a perfect contour and producing a smooth, uniform coat on the lips."

It also saved precious minutes for ladies in a hurry (or their fretful companions). "Many minutes are often required by free hand operation with unsatisfactory or imperfect effects," noted the impatient inventor. "Less than fifteen seconds are required to put a perfect coat of cosmetic on the lips by using the guide."

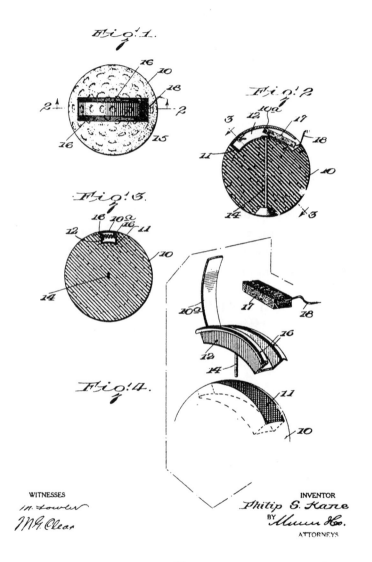

Fig.1.

Fig.2.

Fig.3.

Fig.4.

WITNESSES
M. Fowler
M.G. Clear

INVENTOR
Philip S. Kane
BY
Munn & Co.
ATTORNEYS

GOLF BALL
Patented May 4, 1926
by PHILIP SCHUYLER KANE
Kane, Pennsylvania

No golfer could ever possibly lose a Kane Golf Ball. Like a small bomb, it had a short fuse sticking out that was attached to slow-burning powder underneath the cover. Just light it up, shout "Fore!" and swing away. No matter how deep into the woods it landed, a helpful plume of smoke would mark the ball's location.

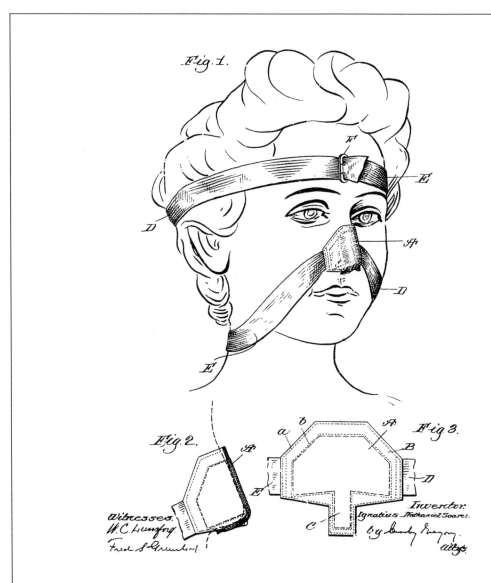

Fig. 1.

Fig. 2.

Fig. 3.

Witnesses.
W. C. Linnford
Fred S. Greenleaf

Inventor:
Ignatius Nathaniel Soares
By Bundy Gregory
Attys.

NOSE-SHAPER
Patented April 23, 1907
by IGNATIUS N. SOARES
Framingham, Massachusetts

Before the nose job, there was the Soares Nose-Shaper.

"The noses of a great many persons are slightly deformed," Mr. Soares stated. "Such deformity can frequently be remedied by a gentle but continuous pressure."

He proposed a nose-shaped piece of sheet metal, lined with chamois and attached to bands that wrapped around the neck and buckled on the forehead. "The result is that a slight but steady tension is produced . . . and the nose is thus gradually but steadily brought into normal shape."

Mr. Soares said his nose shaper could be worn around the clock. But he noted helpfully, "It will usually be found preferable to wear the device at night."

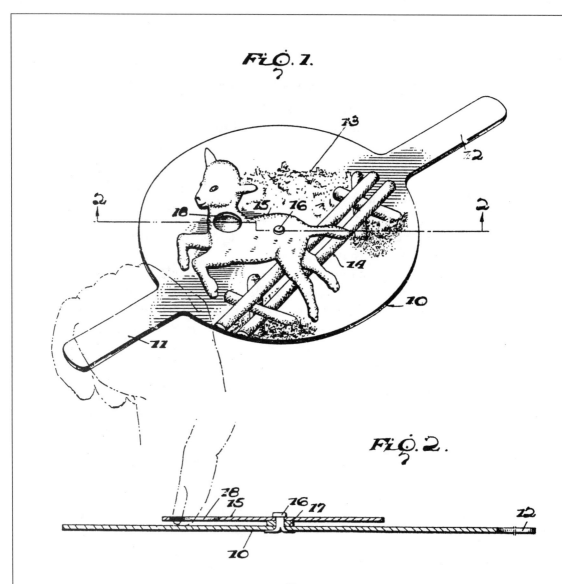

FIG. 1.

FIG. 2.

SLEEP INDUCER
Patented May 19, 1942
by Benjamin K. Greenwood
Pueblo, Colorado

r. Greenwood was a firm believer in counting sheep to chase away the "diversified subjects or details which at the very opportunity of sleep become so intrusive or poignant." The technique had only one flaw: "The effort is all mental and any fatigue produced is also wholly mental."

So he tacked an actual picture of a sheep onto a paddle. Whenever those poignant details intruded, you'd just stick your finger into a hole in the sheep's neck and turn it around in circles. Soon, Mr. Greenwood promised, "muscular effort will be coordinated and synchronized with mental concentration to induce sleep."

Fig. 1

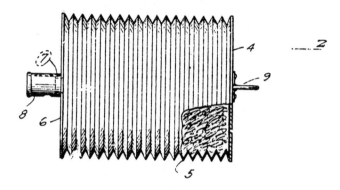

Fig. 2

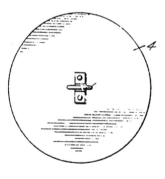

Fig. 3

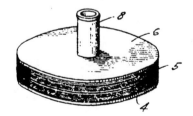

BREATH DETECTOR
Patented February 10, 1925
by George Starr White
Los Angeles, California

"A subject under normal conditions does not smell his own breath," wrote Mr. White, articulating a problem that has undone more than one romantic encounter.

His solution was a small bellows that a person with dubious breath would inflate by blowing several times into a tube. Then he would blast the bellows up his nose, "expel the gases and vapors"—and pray his companion didn't find the whole procedure an even bigger turnoff than bad breath.

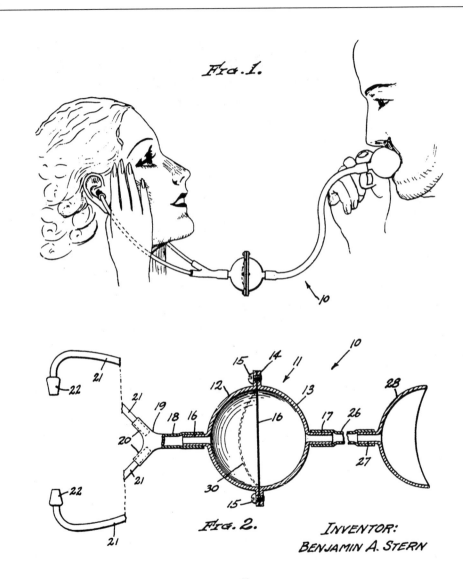

Fig. 1.

Fig. 2.

INVENTOR:
BENJAMIN A. STERN

APPARATUS FOR EXERCISING THE HEARING ORGANS
Patented November 23, 1937
by Benjamin A. Stern
Los Angeles, California

What workout could be truly complete without a hearty session of ear calisthenics?

You stick your Stern Apparatus earplugs into both ears and hand the other end to your training buddy. He blows into the mouthpiece and then starts talking.

"The ear drums of the patient are thus simultaneously subjected to a mild super-atmospheric air pressure from the outside and to normal voice vibrations," declared Mr. Stern, who recommended a good round of earobics "at frequent intervals throughout the day."

Fig. 1.

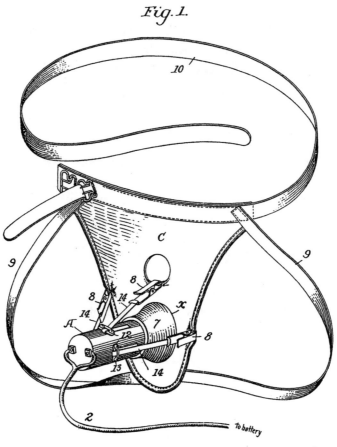

10

C

9   9

8
8   14
14
A   7   x
8
12
13   14

2   to battery

Inventor

Willard F. Main

Foster Freeman & Watson

Witnesses

J. G. Stukel

Jhm Gillman, Jr.

By

Attorneys

SURGICAL APPLIANCE
Patented September 5, 1905
by WILLARD F. MAIN
Iowa City, Iowa

Mr. Main was one of dozens of turn-of-the-century tinkerers who flooded the Patent Office with devices to prevent bed-wetting. His was particularly imaginative.

You strap on a rubber "receptacle." Inside is a battery circuit, interrupted by a "soluble wafer" and wired to an alarm next to your bed. Wet your bed, and the wafer dissolves, the circuit completes, and the alarm goes off.

"In the course of time," Mr. Main promised optimistically, "the habit of awakening as soon as any desire of emission is felt will be established."

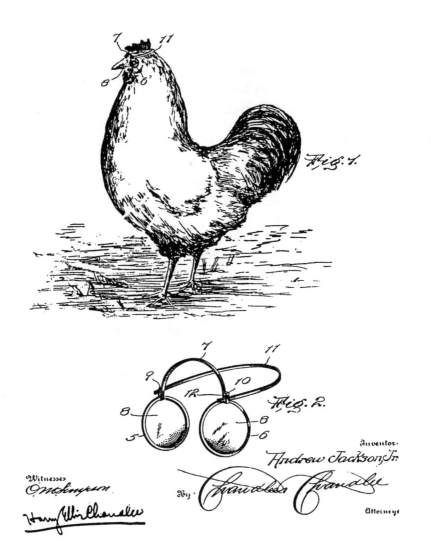

Fig. 1.

Fig. 2.

Inventor.
Andrew Jackson Jr.

Witnesses
Omckingson
Harry Ellis Chandler

By Chandless Chandler

Attorneys

EYE PROTECTOR FOR CHICKENS
Patented June 16, 1903
by ANDREW JACKSON, JR.
Munich, Tennessee

Why put spectacles on a chicken? "So that they may be protected from other fowls that might attempt to peck them," Mr. Jackson wrote.

His chick specs came on an elastic band, with the eyepieces connected on a springy strap. That way they "may be easily and quickly applied and removed," he declared, no doubt to reassure farmers that they wouldn't get pecked either.

(No Model.)

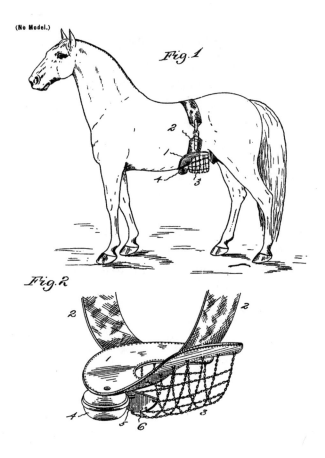

Fig.1

Fig.2

VETERINARY APPLIANCE
Patented September 2, 1902
by NEIL STALKER
West Hartford, Connecticut

Mr. Stalker's appliance was a pouch that strapped on to a stallion "for the purpose of preventing the impairment of his vitality." The inventor explained that "valuable and finely-bred trotting stallions" get "easily excited." This can sap their strength and "thus destroy their value for trotting and breeding purposes."

The solution: Stick a little disk connected to a bell in the pouch. When the stallion reaches "a state of excitement," the disk moves forward and the bell rings, to "notify the hostlers that he should have attendance." What are the hostlers supposed to do at that point? Mr. Stalker didn't say.

Fig.1.
a
a
Fig.2.
a
a
A
B
b
Fig.3.
a
a
A
b
B
Fig.4.
d
c
d
d
c
d
f
e
Witnesses
Edwin G. McKee
Ralph S. Warfield.
Inventor
Leonidas E. Wilson.
by Rhesa G. DuBois & Co.
his Attorneys.

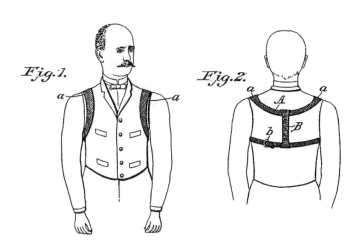

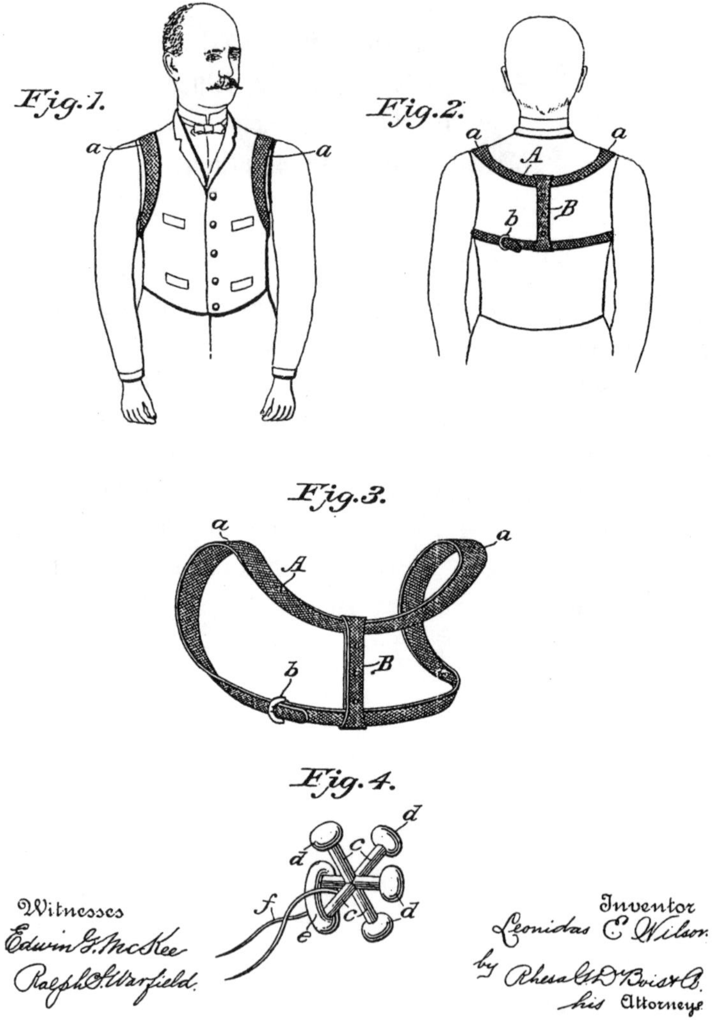

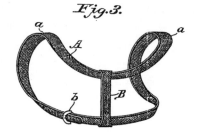

SHOULDER BRACE AND ANTISNORING ATTACHMENT
Patented December 11, 1900
by LEONIDAS E. WILSON
Broken Bow, Nebraska

This hideous little "prodding device" had four metal knobs. It tied to the back of a shoulder harness, "to remind the sleeper in case he lies upon his back by the projections prodding him."

One shortcoming the inventor didn't address: Whoever slept next to the Wilson attachment would surely soon long for the good old days when the wearer merely snored, instead of screaming in pain from all the prodding.

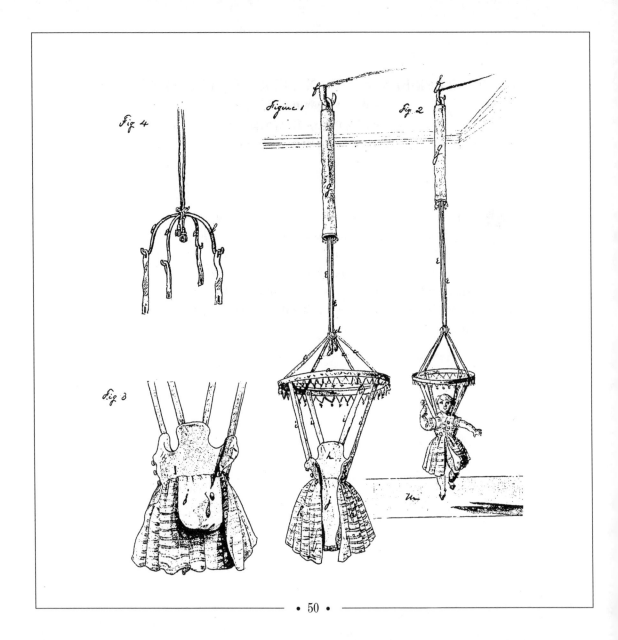

Fig 4

Figure 1

Fig 2

Fig 3

MACHINE FOR EXERCISING CHILDREN
Patented September 4, 1847
by GEORGE W. TUTTLE
New York, New York

Mr. Tuttle dubbed his hoop on a spring a "Baby-Jumper" and swore it was not simply for "amusement."

"It will be found highly useful for exercise to all infants, a great relief to nurses, and useful in giving strength of body and limbs to sick or lame children of a larger growth," he wrote.

Plus a special aesthetic bonus! "Every part of the jumper may be highly ornamented and thus become a beautiful article of furniture."

Fig. 1.

Fig. 2.

Witnesses:
S. H. Wales
J. W. Hamilton

Inventor:
Alpheus Myers.

REMOVING TAPE WORMS
Patented November 14, 1854
by ALPHEUS MYERS
Logansport, Indiana

Mr. Myers was searching for a way to remove tapeworms from the human stomach "without employing medicines, and thereby causing much injury."

His idea of a safer alternative was a three-quarter-inch gold metal "trap," with bait ("any nutritious material") in a little box on a spring. The sufferer would fast (for a "suitable duration to make the worm hungry") and then swallow the trap.

The happy result: "The worm seizes the bait, and its head is caught in the trap, which is then withdrawn from the patient's stomach by the string which has been left hanging from the mouth, dragging after it the whole length of the worm."

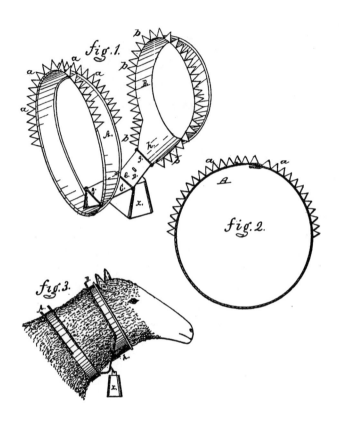

fig.1.

fig.2.

fig.3.

Witnesses
D. C. Allen
A. L. Allen

Inventor
James R. Speer
By J. J. Johnston.
his attorney

SHEEP-PROTECTOR
Patented June 3, 1879
by JAMES R. SPEER
Pittsburgh, Pennsylvania

❦

This device was a small contribution to the arms race in the battle between dog and sheep. It entailed fitting the poor sheep with two metal collars featuring sharp, serrated teeth.

The bloody upshot: "In case the dog does make an attack, the serrated collars will so lacerate its mouth and cause such pain that it will cease its attack."

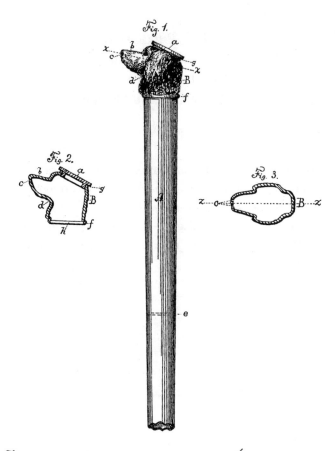

Fig. 1.

Fig. 2.

Fig. 3.

Witnesses:

James J. Richardson.
D. J. Dixon

Inventor:

M. L. Baxter.

SPITTOON CANE
Patented May 3, 1881
by Myron L. Baxter
Aurora, Illinois

❧

"It is, perhaps, deplorable," wrote Mr. Baxter, "but none the less true, that many men who are addicted to the habit of chewing tobacco cannot be happy or contented without it during the continuance of a religious service, a lecture, or other entertainment, and those who are too well bred to spit upon the floor must have much of their enjoyment spoiled or stay away altogether."

So he designed a cunning hollow cane, capped with a dog's head punctured with holes in the nostrils. "The user has only to carelessly insert the projection between his lips—a motion very common among the users of ordinary canes."

Fig. 1.

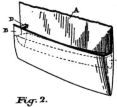

Fig. 2.

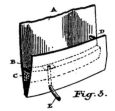

Fig. 3.

Witnesses.

Lewis Tomlinson

F. B. Fetherstonhaugh —

Inventor.

John Maguire

by Donald C. Ridout &c.

Attorneys

WATERPROOF COAT
Patented February 27, 1883
by JOHN MAGUIRE
Toronto, Canada

Mr. Maguire fixated on a flaw of most ordinary raincoats: Water drips from the bottom onto the wearer's legs.

Not his coat. It folded up at the bottom to form a kind of cloth trough—and a little tube off to one side spouted all the water away.

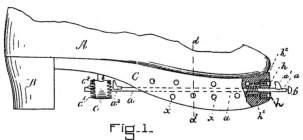

Fig. 1.

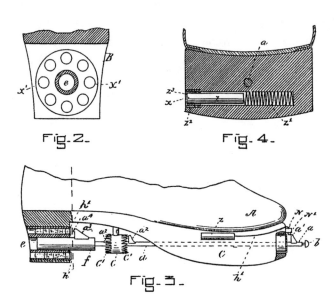

Fig. 2.

Fig. 4.

Fig. 3.

WITNESSES
Bowdoin S. Parker.
Fred Harris.

INVENTOR
John Cassidy
by his atty
Clarke & Raymond.

CLOG OR SHOE
Patented May 15, 1883
by JOHN CASSIDY
Cambridge, Massachusetts

A pioneer in exploding footwear, Mr. Cassidy promised, somewhat darkly, "pleasing and startling effects" from his unusual clog.

The sole came with a cavity for "percussion-caps, torpedoes, or other pyrotechnics." At the front of the sole was a hidden fuse—so you could make a big bang with just a tap of your toe.

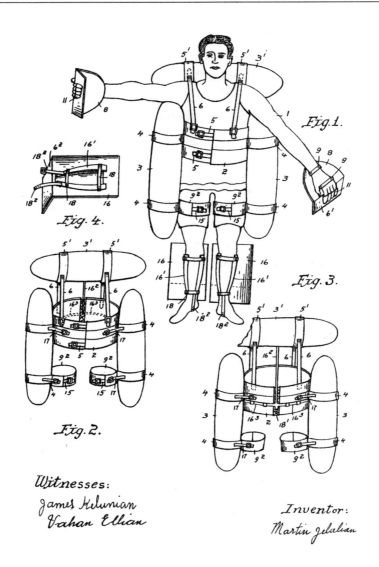

Fig. 1.

Fig. 4.

Fig. 3.

Fig. 2.

Witnesses:
James Kelunian
Vahan Ellian

Inventor:
Martin Jelalian

SWIMMING DEVICE
Patented February 20, 1917
by MARTIN JELALIAN
Cranston, Rhode Island

The point of the Jelalian Swimming Device was to swim "in a most safe, successful and rapid manner." In fact, the gadget would surely have taken so long to strap on that it couldn't have been much of a time-saver.

It featured an air bag on each side of the swimmer's body, a third air bag for the shoulders, a concavo-convex board ("for propelling purposes") on each hand, and two more boards for the legs.

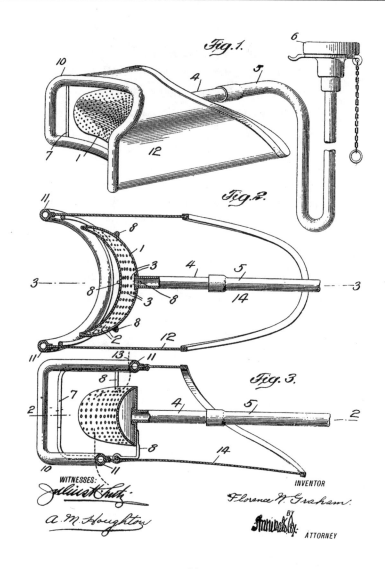

Fig.1.

Fig.2.

Fig.3.

WITNESSES:

INVENTOR

Florence N. Graham.

BY

ATTORNEY

APPARATUS FOR IMPROVING THE CONTOUR AND
CONDITION OF THE CHIN AND THROAT
Patented December 12, 1916
by FLORENCE N. GRAHAM
New York, New York

his early flabbuster was aimed at "the condition often referred to as 'double chin.'" Mrs. Graham was looking for an alternative to manual chin massages, which, she wrote, "usually require the employment of an expert, are not always effective and often result in a flabby or baggy condition."

Her counterproposal: Dump your chin into a little basin and blast it with jets of hot water.

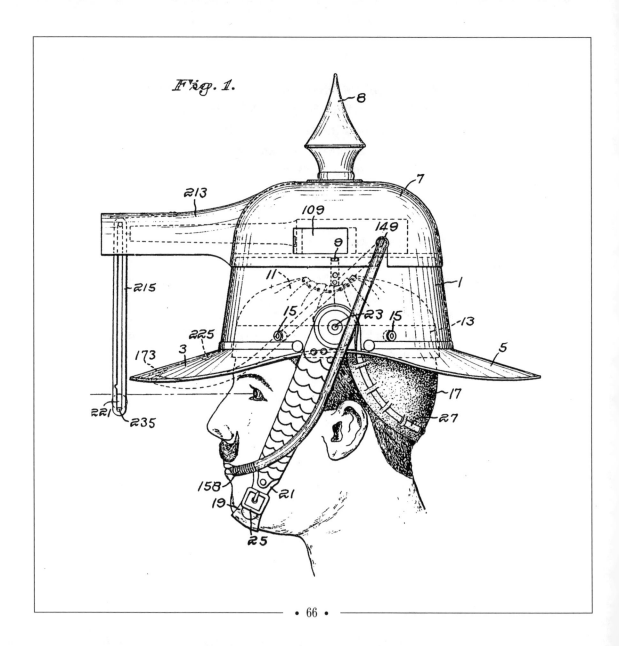

Fig. 1.

WEAPON
Patented May 16, 1916
by ALBERT B. PRATT
Lyndon, Vermont

This terrifying creation would let you fire bullets from the front of your hat by blowing through a little tube in your mouth.

To Mr. Pratt, this arrangement offered several obvious advantages. Your hands and feet would be free to defend yourself by other means. Another plus: "In hunting at night if an animal made a sound in underbrush, the head of the marksman would be instinctively turned in the direction of the sound and then the gun would be fired, without the use of the eyes of the marksman."

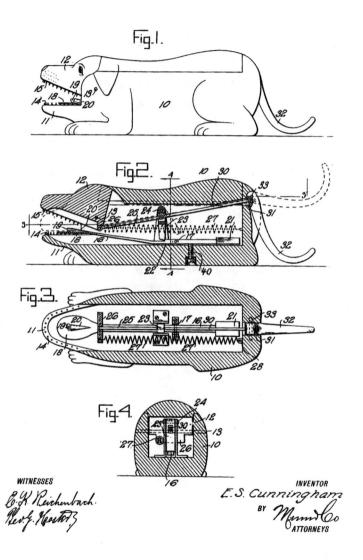

Fig.1.

Fig.2.

Fig.3.

Fig.4.

INVENTOR
E. S. Cunningham
BY
Munn & Co
ATTORNEYS

ANIMAL TRAP
Patented April 25, 1916
by EDWIN SCOTT CUNNINGHAM
Mansfield, Illinois

M r. Cunningham took pains to explain the unique virtues of his trap: You could set it without hurting your fingers, release the dead animal without touching it, and keep the animal from actually consuming the bait.

He never explained just why he designed the trap "in the outline of a dog or other animal"—whose jaws would probably be the last place in the world any animal would innocently wander into.

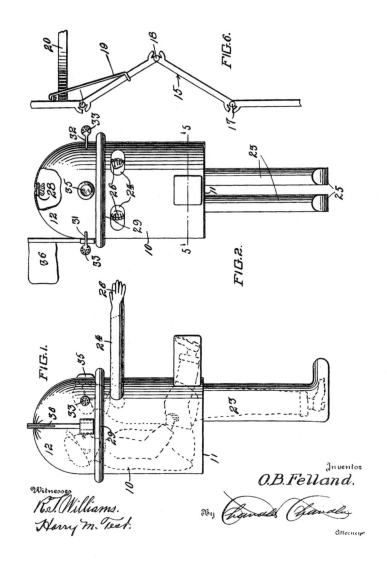

FIG. 6.

FIG. 2.

FIG. 1.

Witnesses
R.T. Williams.
Harry M. Test.

Inventor
O.B. Felland.

By

Attorney

LIFE BUOY
Patented August 3, 1915
by Olaus B. Felland
Edmonton, Canada

Using the Felland Life Buoy, a passenger on a sinking ship would leap into a little hoop, pull a waterproof cloak over his head, and plunge safely into the ocean.

Inside the envelope he'd find everything he needed while he waited for his rescue: an inflatable tube, a glass porthole, a signal flag, leg holes for kicking, and a system of bulbs and funnels "designed to exhaust the foul air from the contraption."

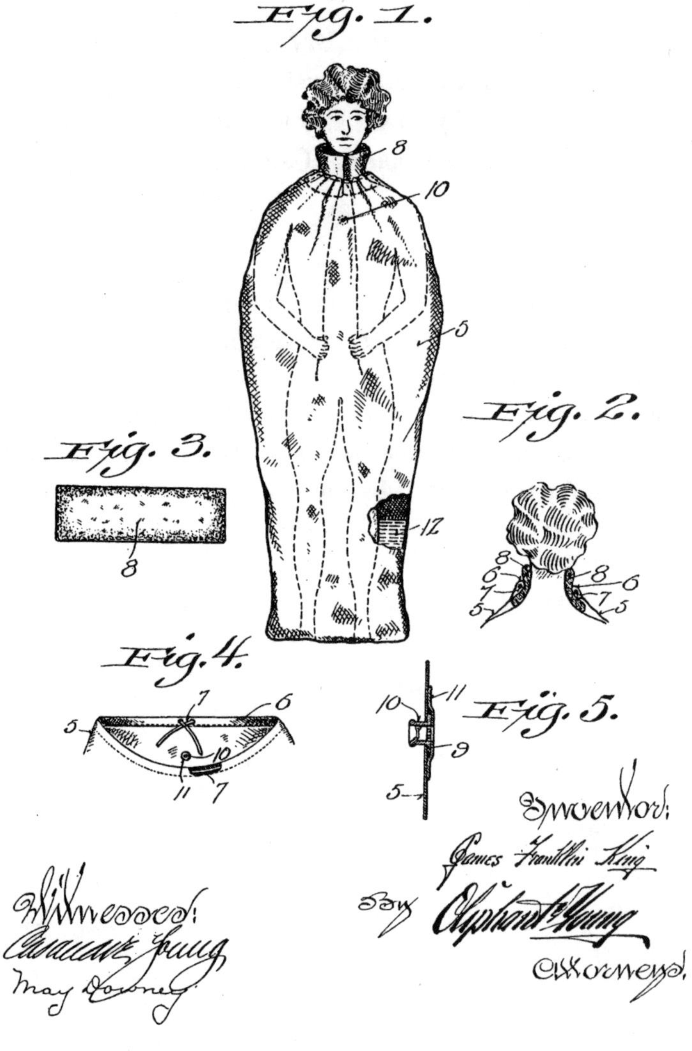

Fig. 1.

Fig. 2.

Fig. 3.

Fig. 4.

Fig. 5.

BATH APPARATUS
Patented July 1, 1913
by JAMES FRANKLIN KING
Milwaukee, Wisconsin

The portable bath was a nifty way to keep a traveler clean "where the ordinary bath conveniences are not available." The bather would climb into a waterproof bag and pull a drawstring to tighten a wool collar around his neck. Then an attendant would inject "air, steam, or other aeriform fluid" into the bag.

Mr. King soberly assured the patent examiners his invention would offer pleasure as well as hygiene: "By alternate crouching and rising in the bag or by rolling with the same upon a bed or floor on the part of the bather, [the liquid] may be made to surge in simulation of sea waves and thus afford gratification to said bather."

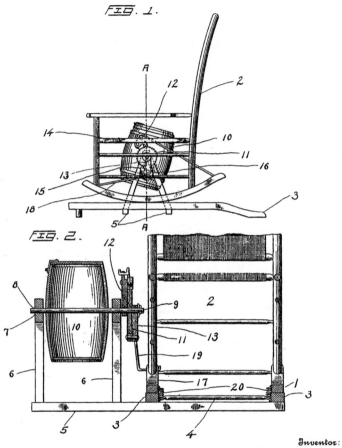

FIG. 1.

FIG. 2.

Inventor:
Alfred Clark

CHURN
Patented January 28, 1913
by ALFRED CLARK
East Corinth, Maine

❧

Mr. Clark attacked the problem of how to harness all that energy he generated by rocking back and forth in his rocking chair. Answer: Connect the chair to a butter churn with a shaft, a ratchet wheel, some rods and pawls, and a dozen other parts.

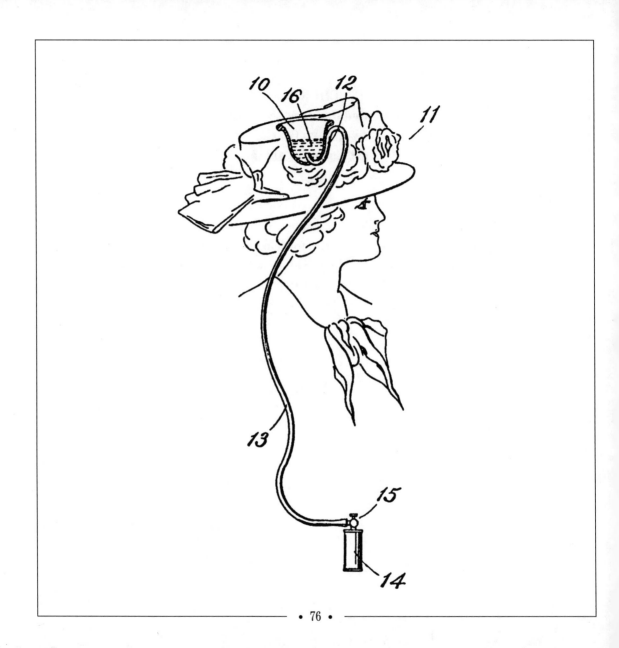

BUBBLE HAT
Patented March 26, 1912
by ALDEN L. McMURTRY
Sound Beach, Connecticut

r. McMurtry envisioned nestling a cup partly filled with soapy water in a woman's hat. A tube would connect it to "a small portable gas tank" filled with pressurized hydrogen, tucked away in a pocket or other hiding place.

Why a bubble hat? "A number of persons such as the members of a chorus may be equipped with such apparatus so that they may together or singly cause bubbles to arise from their heads, either in conjunction with a song or otherwise," Mr. McMurtry wrote.

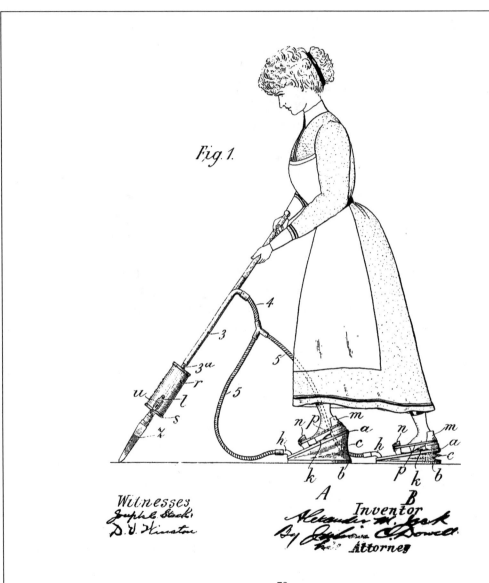

Fig. 1.

Witnesses
Joseph E. Stack
D. V. Winston

Inventor
Alexander W. Cook
By Joseph C. Dowell
Attorney

• 78 •

PORTABLE AIR WITHDRAWING OR VACUUM PRODUCING APPARATUS
Patented February 6, 1912
by ALEXANDER MACKENZIE JACK
Sheffield, England

Cleaning the house is never fun, but it was an absolute nightmare with turn-of-the-century vacuum cleaners. The suction came from an elaborate bellows, operated by pedal or hand. A pair of housemaids had to drag it around from room to room, pumping furiously.

Mr. Jack ingeniously proposed affixing the bellows to the housemaid's feet. That, he wrote, would let her clean the house "with a minimum of muscular fatigue."

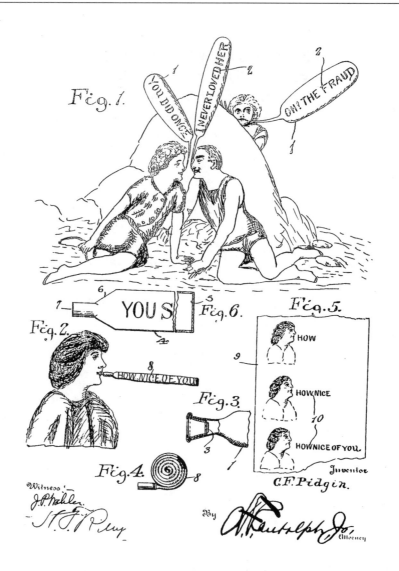

MOTION PICTURE AND METHOD OF PRODUCING THE SAME
Patented September 18, 1917
by Charles F. Pidgin
Winthrop, Massachusetts

Mr. Pidgin was a silent-movie fan who was annoyed by the way screens of text kept breaking up the action. So he devised a new technology that would "enable the dramatic situation to be explained or other information to be conveyed simultaneously with the action and without the use of a separate screen or picture or marginal sign or inscription."

The sound track? Nope. The rolled-up New Year's Eve–style inflatable tube, with text printed on the side.

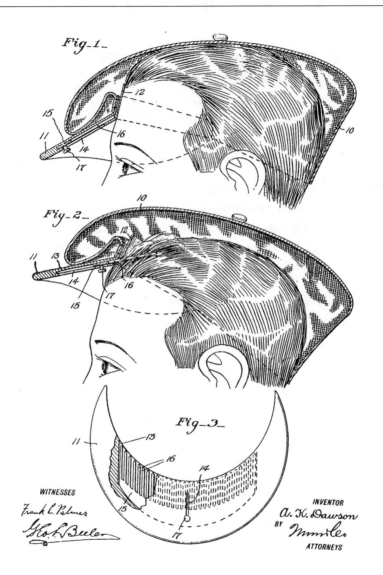

Fig_1_

Fig_2_

Fig_3_

WITNESSES

Frank C. Palmer

Geo. L. Beeler

INVENTOR
A. K. Dawson
BY
Mumile.
ATTORNEYS

COMBINED HEAD COVERING AND HAIR COMB
Patented January 6, 1920
by Alva K. Dawson
Jacksonville, Florida

The Dawson comb-hat was intended to alleviate a common fashion problem for young men returning from a vigorous day of golfing or motoring: the disheveled pompadour.

It cunningly concealed a little sliding comb beneath the visor. Thus the suave owner could "comb up his hair coincidentally with the removal of the head covering from his head, and hence without rendering himself conspicuous in so doing."

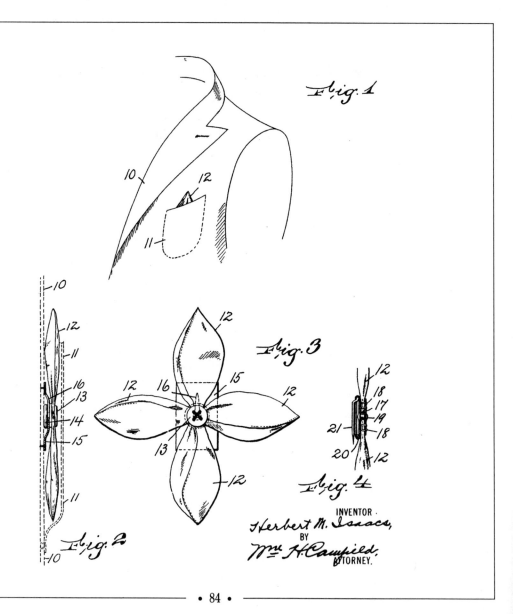

Fig. 1

Fig. 2

Fig. 3

Fig. 4

INVENTOR
Herbert M. Isaacs,
BY
Wm H. Camfield,
ATTORNEY.

POCKET ORNAMENT
Patented December 7, 1920
by HERBERT M. ISAACS
Newark, New Jersey

For the well-dressed cheapskate, Mr. Isaacs offered an alternative to pocket handkerchiefs. Like so many inventors, he seized upon an economic market flaw that everyone else had previously ignored: You pay for an entire handkerchief, but you show only a small corner of it.

"This is, in a sense, expensive," he observed. So he proposed a thrifty pinwheel of four handkerchief corners. You slipped it in your pocket and selected one corner to display. (In fact it was doubly thrifty, he hastened to point out, "because remnants in many cases can be used in making up the device.")

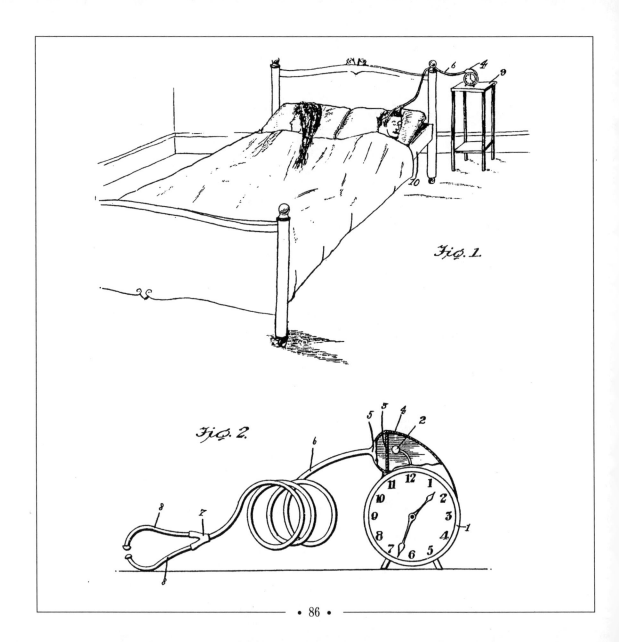

Fig. 1.

Fig. 2.

ALARM CLOCK
Patented February 1, 1921
by CHARLES W. WALLER
Chicago, Illinois

❀

The marriage bed was a happy place with a Waller Alarm Clock on the night table. The lady of the house could sleep as long as she liked. But the gentleman would never be late for work because he slept with an alarm conveniently wired into his ears.

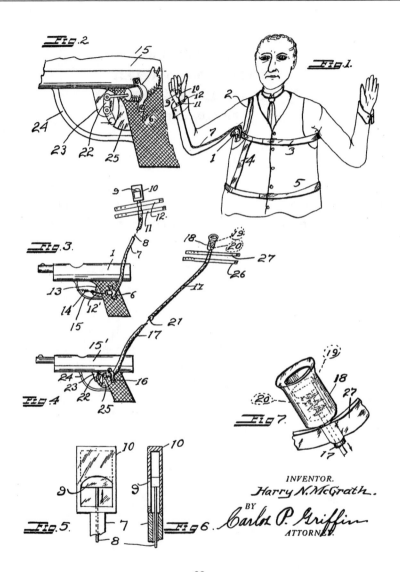

Fig.2

Fig.1

Fig.3

Fig.4

Fig.7

Fig.5

Fig.6

INVENTOR.
Harry N. McGrath.
BY
Carlos P. Griffin
ATTORNEY.

GUN FIRING DEVICE
Patented May 3, 1921
by HARRY N. MCGRATH
San Francisco, California

Mr. McGrath knew how stickups worked. "The robber usually demands that the cashier throw up his hands thus preventing him from firing a pistol."

Not a McGrath pistol. It poked out from the cashier's armpit. A wire up his sleeve connected to a remote-control trigger in the palm of his hand.

Couldn't a careless cashier make a slip of the fingers while counting bills and blow his arm off by mistake? Mr. McGrath thought of that: "In order to make the gun perfectly safe, a blank cartridge can be placed in the magazine to be fired first, followed by a ball cartridge."

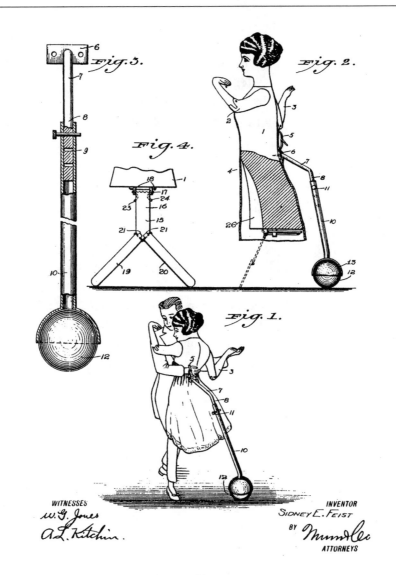

Fig. 3.

Fig. 2.

Fig. 4.

Fig. 1.

INVENTOR
SIDNEY E. FEIST
BY
Munn & Co
ATTORNEYS

FIGURE FOR BALLROOM DANCING PRACTICE
Patented May 17, 1921
by SIDNEY E. FEIST
Brooklyn, New York

Mr. Feist designed the perfect ballroom partner. It had no feet to step on, no awkward chitchat, and a little strap on the back to clamp your hand in place.

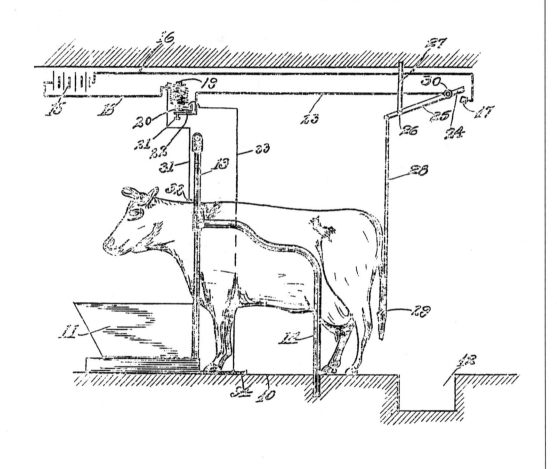

ELECTRICAL SYSTEM FOR USE IN CATTLE STALLS
Patented April 4, 1922
by Elmer Swensen
Valders, Wisconsin

In the annals of truly sadistic agricultural ingenuity, Elmer Swensen's place will always be secure. His mission, reasonable enough, was to keep cattle stalls clean by encouraging cows to drop their waste in a gutter behind them.

When the unfortunate cow lifted her tail, it raised a rope, which moved a lever, which flipped a switch, which sent an electrical current through a bar across the animal's shoulders and also through a plate under her front legs.

"The animal will, therefore, upon raising its tail and closing the switch, be subjected to a sudden electrical shock and step backwardly off the mat to a proper position to deposit the droppings in the gutter."

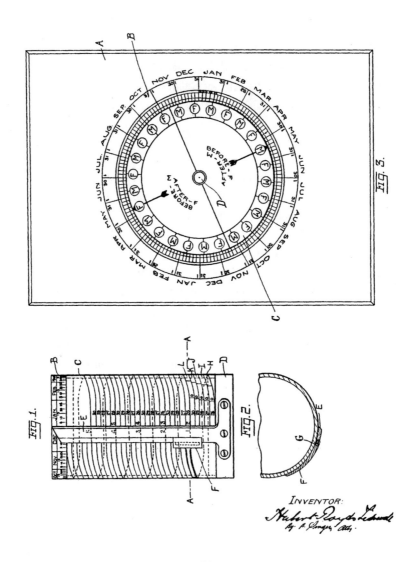

FIG. 3.

FIG. 1.

FIG. 2.

INVENTOR:

Hubert Lloyd Ledwidge
By P. Singer, Atty.

SEX CALCULATOR
Patented September 5, 1920
by HUBERT R. TIDSWELL
Winnington, England

This handy cardboard wheel performed an extremely useful calculation for expectant parents. Dial up the due date, and it predicted the baby's sex. You could also calculate the proper date for "fertilization" if you wanted to breed a particular sex.

"This invention," Mr. Tidswell explained to the patent examiners, "is based on the well-known and proved theory that in normal women the ovaries ovulate alternately, and that one ovary always produces male ova and the other female ova." The Patent Office pondered that explanation and then gave Mr. Tidswell patent number 1,428,065.

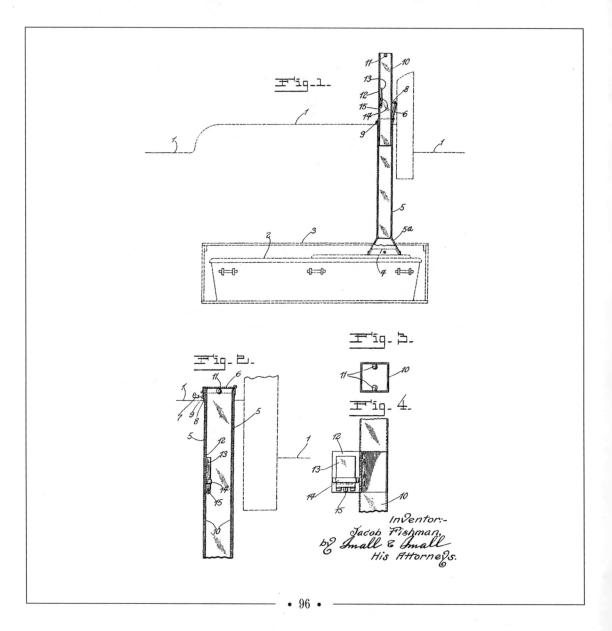

Fig. 1.

Fig. 2.

Fig. 3.

Fig. 4.

Inventor:-
Jacob Fishman,
by Small & Small
His Attorneys.

ATTACHMENT FOR CASKETS
Patented November 28, 1922
by JACOB FISHMAN
St. Louis, Missouri

For Mr. Fishman, a traditional casket had a significant limitation: Once you buried it in the ground, there was no way to see the corpse.

The Fishman casket overcame this deficiency. It had a telescoping tube you could pull out of the ground. Turn on a light, peep inside, and there's Grandma.

The discreet inventor designed a lock for the cover, "to prevent the admission of rain and particles of earth and to insure that unauthorized persons may not indulge their curiosity."

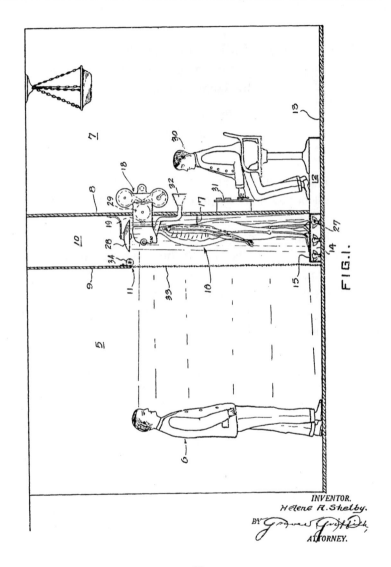

FIG.I.

INVENTOR.
Helene A.Shelby.

BY *[signature]*
ATTORNEY.

APPARATUS FOR OBTAINING CRIMINAL CONFESSIONS AND PHOTOGRAPHICALLY RECORDING THEM
Patented March 4, 1930
by HELENE A. SHELBY
Oakland, California

Helene Shelby knew what it took to make a criminal crack. Her invention would "produce, when operated, a state of mind calculated to cause him, if guilty, to make confession thereof."

Specifically, as the interrogation was under way in a dark room, a curtain would rise and reveal the figure of a skeleton. Flashing lights in its eyes would "add to the mystification." The result: The criminal would break down and confess, while a hidden camera in the skull recorded the whole thing.

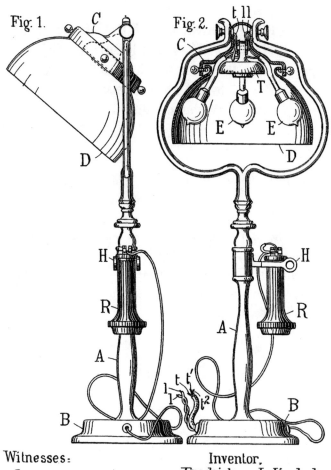

Fig. 1.

Fig. 2.

COMBINED TELEPHONE AND ELECTRIC LAMP
Patented July 4, 1911
by FREDRICK J. KERBEL
New York, New York

What do you get when you cross Thomas Edison and Alexander Graham Bell? Fredrick J. Kerbel! (Just speak clearly into the lampshade.)